WHAT DOES YOUR DATA SAY?

*A Conversation–Centric Design™
Primer for Power BI Content
Development,
Version 2*

Treb Gatte

Marquee Insights

ISBN- 978-1-7333418-2-0

Cover design by: Naomi Moneypenny

Printed in the United States of America

"A prudent question is one-half of wisdom."
—FRANCIS BACON

Contents

CHAPTER 1. INTRODUCTION TO CONVERSATION-CENTRIC DESIGN™

Many organizations are challenged when beginning their business intelligence (BI) journey. Many will jump into designing their first BI content with no clear framework for design. This can lead to many undesirable outcomes.

They also struggle as to where they need to invest in creating content. Usually, the project sponsor will scope the project to address several transactional needs without regard to the bigger organizational need. This creates content sprawl over time and increases maintenance costs. Most projects focus on the needs of Middle Managers as a result.

Figure 1 Common focus of Business Intelligence Projects

Also, it can be difficult for managers to make prioritization decisions when the choices are presented in terms of the technology used, rather than the business need.

This book introduces the Conversation–Centric Design approach that ensures created BI content is aligned to the business need, structured to help you manage scope and ensures client clarity as to where the content should be used. This process will also help you prepare for Copilot for Power BI if this capability is in your future.

We typically see the following happen in many organizations. Later in the book, we will discuss how to address these situations.

Just give me the spreadsheet

Steve closed the Microsoft Teams call where he had a meeting with the Finance team to discuss their report design. He felt crushed and needed a break to clear his mind. He came back to his desk and pinged Dipti, his team manager, over Teams. She answered and saw his face. She could tell something was wrong. "Hey Steve, how did the meeting go?" she asked with concern. Steve shook his head and said, "It was terrible, Dipti. All the users wanted was a web page with a spreadsheet on it. They want to see all the details."

Dipti was confused as she had talked to the Finance leadership who had complained about the constant fire drills and how it took them 3 days to get any answer to their questions. Their team always had to manually create the answer from scratch. Dipti had shown them how Power BI could help them get answers to business questions quickly.

"Didn't you show them the dashboard designs we had worked on?" she asked. He replied, "I did, but they didn't care. They could not see how it applied to them. I asked them what questions they were trying to answer from a spreadsheet. They could not explain why they needed the details. They kept insisting they needed to see all 50,000 rows of data."

Dipti frowned. "We must try again with this group. Their leadership wants less chaos when questions are asked. Otherwise, this project will not meet expectations and put our jobs at risk."

Tower of Babel problem

Dipti was eagerly waiting for Denise, an external BI consultant, to join the call. Denise was a Microsoft MVP (Most Valuable Professional) and had a lot of experience in running large BI projects for one of the industry competitors. Dipti wanted to know how the walkthrough of the PMO (Project Management Office) reporting with five Corporate Vice Presidents (CVP) had gone.

"Hi, Denise! I'm glad you could make it." Dipti greeted her warmly as Denise appeared on the screen. "Hi, Dipti," Denise said with a serious tone. "I have some bad news for you. The walkthrough was a disaster when we got to the Project Health report." "Oh no!" Dipti exclaimed. "What happened?"

Denise sighed and recounted the events. "It all started when I showed them the Project Health indicator. They all challenged why a certain project was marked as Red. I explained to them the logic that we got from the primary stakeholder, of how we determined the Red-Amber-Green status. That's when they looked at each other and said, 'That's not how we define it.' It turned out that each CVP had a different definition of what Project Health meant!" Denise said with frustration. "It was like the Tower of Babel story. They all started with the same metric but ended up with different interpretations that made no sense to each other."

Dipti felt a pang of anxiety. "So, what's the impact?" Denise said, "We have to review the definitions with them this Thursday, try to reach a consensus across the five groups, and update the data models accordingly. Our roll-out is delayed for sure, but I don't know by how much yet." Dipti leaned back in her chair and wondered, "How did we miss this?"

Why are you working on that?

Dipti was presenting the monthly project status update to Heather, her CVP. She was presenting a slide deck showing the progress and challenges of the project. "As you can see, we have completed 80% of the tasks in the project plan, and we are on track to meet the deadline for the PMO reporting," said Dipti, pointing to a Gantt chart on the screen. "However, we have encountered some issues with the finance reporting, which is one of the key deliverables of the project."

Heather looked at Dipti with a puzzled expression. She had expected to hear more about the PMO reporting, which was her main priority and concern. She interrupted Dipti and asked, "What do you mean by finance reporting? What is the scope and purpose of that deliverable?"

Dipti was surprised by Heather's question. She thought she had explained the finance reporting clearly in the previous updates. She replied, "The finance reporting is part of the project scope that we agreed on at the beginning of the project. The dashboards will show the financial performance and projections of our business unit, helping us make better decisions about our budget allocations. My update last month covered the data model and report prototype work that is supporting the financial reporting deliverable."

Heather shook her head, "I don't recall agreeing to that deliverable. I don't see how it is relevant or useful for our business unit. We already have a finance team that handles all the financial reporting and analysis. Why do we need to duplicate their work? Also, it wasn't clear to me that the model work was related to the finance deliverable."

Dipti realized that there was a misalignment between her and Heather. She said, "I apologize for the confusion, Heather. I thought you were aware of the finance reporting deliverable. I remember you mentioned in one of our meetings that you wanted to have more visibility and control over our financial situation. That's why I proposed to include the finance reporting as part of the project scope, and you agreed to it."

Heather thought for a moment, "I think I remember that meeting, but I didn't realize that you were talking about creating a whole new set of dashboards for finance reporting. I thought you were just going to use some existing reports or tools from the finance team. How much time and effort have you spent on this deliverable?"

Dipti checked her notes, "We have spent about 40% of our total project hours on the finance reporting deliverable. We have faced some significant challenges with data quality, integration, and visualization. We have also received a lot of feedback and requests from the target audience, who are mostly senior managers and directors in our business unit. They have expressed a high level of interest and satisfaction with the deliverable."

Heather frowned, "That's a lot of time and effort for a low priority deliverable. I'm concerned that we are not focusing enough on the PMO reporting, which is the most important and urgent project deliverable. The PMO reporting is what will show our value and impact to the executive leadership. How are we doing on that deliverable?"

Dipti felt a surge of anxiety as she realized that she had misjudged Heather's priorities and expectations. She replied, "We are doing well on the PMO reporting deliverable, but we still have some work left to do. We need to finalize the data sources, validate the calculations, test the functionality, and polish the design of the PMO dashboards. We also need to prepare a presentation and a report to showcase our findings and recommendations."

Heather asked, "How much time do you need to complete all that work?"

Dipti stated, "We need about two more weeks to finish everything."

Frowning, Heather grumbled, "That's cutting it very close to the deadline. I'm not comfortable with that risk level. I want you to prioritize the PMO reporting over everything else. Let's put the finance reporting on hold until we finish the PMO reporting."

Dipti felt conflicted as she listened to Heather. She knew that Heather had the final say on the project scope and priorities. But she also knew that the team had invested a lot of time and effort on the finance reporting deliverable, and that it had generated a lot of positive feedback and value for their target audience. She wondered if there was a way to recover the situation by satisfying both Heather's needs and those of Finance.

Dipti suggested, "I understand your concern, Heather, but I'm not sure if putting the finance reporting on hold is a good idea. It's almost done, and it has already delivered some tangible benefits for our business unit. It has helped us identify some opportunities for cost savings, revenue growth, and risk mitigation. It has also increased our credibility and trust with the senior managers and directors, who are our key stakeholders and influencers. If we put it on hold, we might lose some of the momentum and goodwill that we have built with them."

Heather reacted, "I'm sure the team has done great work here, but I feel these resources would be better dedicated to our main priorities. The PMO reporting deliverable is what matters the most for our project success as it shows our alignment and contribution to the organization's strategic goals and priorities. Those factors ensure our future funding so we cannot afford to negatively impact it. The finance reporting deliverable is not a core part of our project scope and success. I feel it's a nice-to-have, not a must-have at this point as it's not what we are being evaluated on."

Dipti realized that she could not persuade Heather to change her mind. She conceded, "OK, Heather, I'll have the team prioritize the PMO reporting deliverable and I'll communicate this realignment to Finance. If I get pushback, I'll loop you into the thread for your support. Lastly, I'd like to have finance reporting reconsidered once we finish the PMO reporting. I think it is a valuable deliverable that can help us improve our financial performance and decision making."

Heather nodded, "Thank you for your cooperation and teamwork, Dipti. I appreciate your feedback and suggestions. I'll keep an open mind about the finance reporting. In the meantime, I need the team focused on the PMO reporting and to make sure we deliver it on time and with high quality."

Dipti concurred, "I understand, Heather. we will deliver the PMO reporting deliverable as soon as possible."

Heather directed, "Thank you, Dipti. In the future, I need you to have greater clarity about the scope and purpose of each deliverable in your project updates. Show me how your project activities and priorities align with my expectations and directions. Lastly, keep me informed and involved in any major decisions or changes in your project."

Dipti acknowledged, "Heather, I agree, and I apologize for any confusion or misunderstanding. I will be clearer about the scope and purpose of each deliverable in my future project updates."

Heather said, "Thank you, Dipti. I appreciate your professionalism and commitment. I look forward to seeing a demo of the PMO reporting deliverable soon."

What's really going on here?

Many people who start using Power BI for their business intelligence (BI) needs face a common challenge: how to design effective and meaningful BI content from scratch, with no supporting approach or framework. It is not easy to create reports and dashboards that can help you make better decisions without some guidance and inspiration.

That's why using a blank canvas approach can be frustrating and inefficient. You might end up with a lot of data, but not a lot of insights. Instead of throwing paint at the canvas like the artist Jackson Pollack, you need a more structured and systematic way to plan and develop your BI content. You need to identify the key questions you want to answer, the data sources you need to access, and the best ways to visualize and communicate your findings.

What follows is an approach that we've developed over the last decade helping customers like Kaiser Permanente, HDR, and Pandora Media be successful in their BI journeys. It has been instrumental in helping managers understand clearly how the BI work will add value, enabling them to better prioritize the work. It helped the development team understand and uncover potential definitional issues early in the development process. Lastly, it's helped reduce the instances where the users want a spreadsheet.

Having a clear line of sight to what BI content will be used where and what exactly it will provide makes it easier for end users to adopt.

Figure 2 Artist, Jackson Pollack at work.

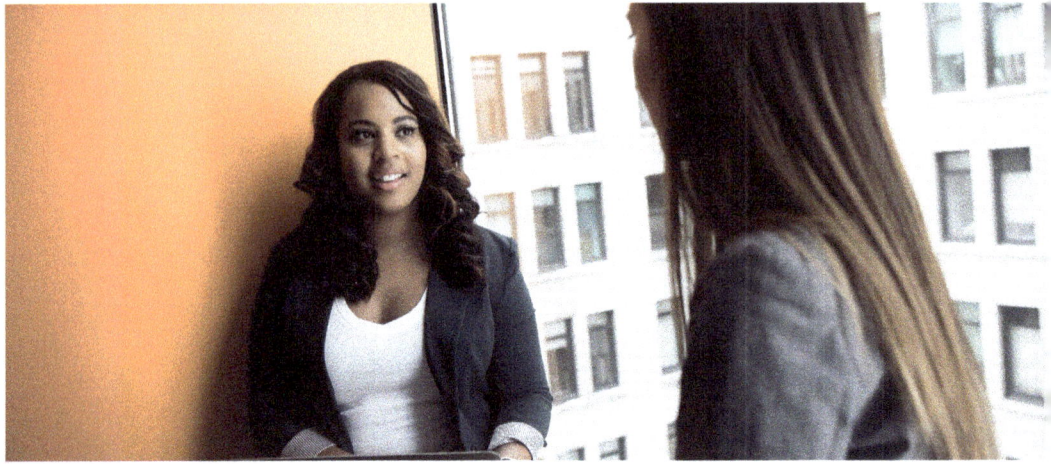

CHAPTER 2. CONVERSATION-CENTRIC DESIGN™ OVERVIEW

The Conversation-Centric Design™ (CCD) process helps you create Business Intelligence content that meets the needs and expectations of your users. It uses your users' daily social interactions that happen in the work environment as a source of inspiration and guidance for your BI content design.

To be clear, the "conversations" to which we're referring above are business discussions that require the types of data and insights that Power BI can provide. In particular, we focus on regularly occurring and business-critical meetings.

Using these conversations creates a framework allowing you to map the work you are undertaking with the situations your users understand, enabling them to better prioritize and understand the work.

The DELTA information needs process

The CCD process leverages the DELTA information needs process to determine if reporting is needed.

- **D**etermine if a situation exists that requires a change
- **E**xamine the relevant data
- **L**earn from the data investigation
- **T**heorize actions to take
- **A**ct and monitor

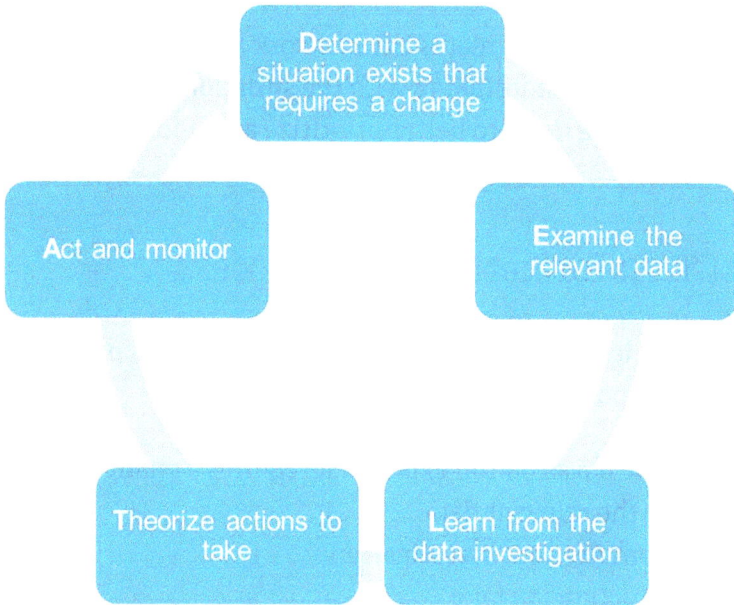

Figure 3 The DELTA Process

These interactions can be formal or informal, such as meetings, calls, emails, chats, etc. The CCD process focuses on formal interactions, which are more structured and predictable. They have a fixed schedule, a defined participant list, and a clear agenda. Therefore, we recommend that you use the CCD process with these formal interactions as a starting point for your BI content design.

By using the CCD process, you can decide which conversations to invest in from a business intelligence perspective. You can identify the most important and relevant conversations for your business goals and priorities. You can also evaluate the impact and value of each conversation for your stakeholders and decision makers.

Why formal interactions?

Using formal interactions makes it clear where the new functionality will contribute value to the organization. Also, it enables successful adoption of new BI content. Users of the new content will be clear about where and how to use it in their work.

Process Overview

The Conversation-Centric Design™ process has four distinct phases as shown in the figure below. The intent is to take you from a position of endless possibilities to a starting point with clear priorities and steps to deliver them.

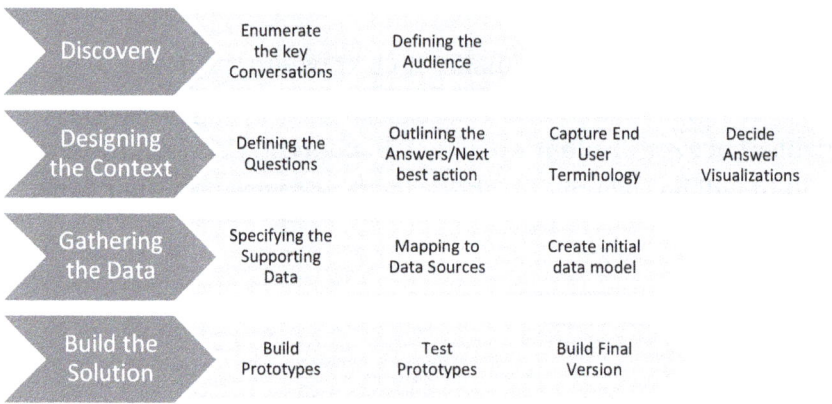

Discovery	Enumerate the key Conversations	Defining the Audience		
Designing the Context	Defining the Questions	Outlining the Answers/Next best action	Capture End User Terminology	Decide Answer Visualizations
Gathering the Data	Specifying the Supporting Data	Mapping to Data Sources	Create initial data model	
Build the Solution	Build Prototypes	Test Prototypes	Build Final Version	

Figure 4 Detailed Conversation Centric Design process

Discovery

Discovery	Enumerate the key Conversations	Defining the Audience		
Designing the Context	Defining the Questions	Outlining the Answers/Next best action	Capture End User Terminology	Decide Answer Visualizations
Gathering the Data	Specifying the Supporting Data	Mapping to Data Sources	Create initial data model	
Build the Solution	Build Prototypes	Test Prototypes	Build Final Version	

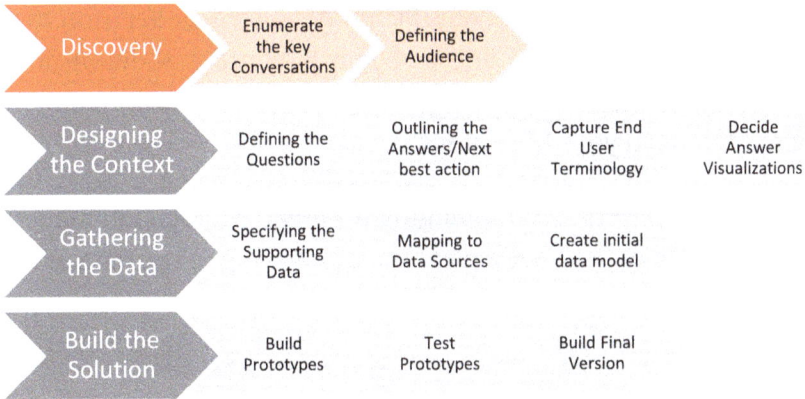

The discovery phase supplies the steps necessary to focus your efforts on key conversations. Starting well drives later success so effort spent here can create large dividends later.

Enumerate the key conversations

First, we start by cataloguing the key conversations. As stated previously, formal interactions like standing meetings are an easy place to start. The goals are to create:

- A recognizable list of conversations
- A list of audience members for each conversation

These design artifacts are used for four purposes, scoping, prioritization, status, and audience definition.

Scoping

One of the benefits of using the CCD process is that it helps you scope your BI content effectively. Scoping is a critical decision that affects the quality and usefulness of your BI content. It determines what data sources, metrics, calculations, and visualizations you need to include in your BI content.

However, scoping can be challenging for business decision makers (BDMs), who may not have a clear understanding of the data work involved and the business value delivered by each BI content element.

The CCD process solves this problem by using a known framing event as the scoping mechanism. A framing event is a formal interaction that has a specific purpose and outcome, such as a monthly review meeting, a quarterly report presentation, or an annual budget planning session. The framing event provides the BDM with a context that helps them understand what the data investment impacts and the relative importance of the event.

For example, if the framing event is a monthly review meeting, the BDM can see how the BI content supports the discussion and decision making in that meeting. They can also compare the importance of that meeting with other framing events in their work calendar.

Using the framing event as the scoping mechanism enables the BDM to make an informed scoping decision. They can prioritize the BI content elements that are most relevant and useful for the framing event. They can also avoid spending too much time and effort on BI content elements that are not aligned with or required for the framing event. This way, they can optimize their data investment and ensure that their BI content delivers maximum value for their business objectives.

Prioritization

The CCD process helps you to decide which conversations to address first, based on their importance and urgency for your business goals and stakeholders. By prioritizing the conversations that have the most impact and value for your business, you can optimize your data investment and ensure that your BI content meets the needs of your key decision makers and influencers.

For example, if your work includes client conversations and internal team conversation to be supported by reporting, management may conclude that client conversations take priority over internal needs. The information needed to make this determination is concise as you've used the conversation to frame what conversations the BI content will support. When you have multiple potential conversations supported by a piece of BI content, you may need to be more granular to determine relative priority of work items. Lastly, if you want to win support from the executive suite, you should prioritize the conversations that involve them and address their pain points and expectations.

Status

The CCD process helps you to track and communicate the progress and challenges of your BI content design, using the formal interactions as a context. Business decision makers (BDMs) are usually not data experts, so they may not understand or appreciate the technical details and complexities of data work, such as data models, measures, and ETL (Extract, Transform and Load) processes. By providing updates relative to the formal interactions, you can help the BDMs to understand how your data work supports and affects their business objectives and outcomes. You can also use formal interactions as an opportunity to get feedback and approval from the BDMs on your BI content design.

Defining the Audience

The CCD process helps you to identify and engage with the potential users and stakeholders of your BI content, using the formal interactions as a source of information and validation. You can capture the audience for each conversation and use it to define your security model and testing groups later in the process.

You can also use it to report status and solicit feedback from the relevant parties. Many business intelligence projects fail to consider the needs and expectations of all the possible users and stakeholders of their BI content. This can lead to dissatisfaction, confusion, or resistance when the BI content is deployed more widely.

Using the STAR model below, you can consider or deliberately exclude the needs of various potential users and stakeholders. Doing so ensures that you design your BI content with a broader perspective and avoid costly rework.

The STAR model is a framework that helps you to identify and analyze the users and stakeholders of your BI content. It identifies 5 key constituent groups, as classified by the following four groups:

Stakeholders and direct managers.

These are the people who will use the BI content to synthesize other conclusions and observations to communicate up. Ensuring that the content provides necessary context to provided information will make their reuse easier and more likely.

Team

These are the people on your team that can use or benefit from the content directly. They may be analysts or specialists who have operational or tactical tasks and responsibilities for your business unit or organization. Typically, they collaborate with the requestor, so it is helpful to get their design input and testing assistance.

Allies and Peers

They are usually peers or colleagues who have complementary or related roles or functions for your business unit or organization. If the requestor has a BI need, it is likely this group has similar needs as well. Their input and expertise should be consulted.

Requestor

This is the person or group requesting the BI content. They are the most vested in the outcome and should be the focal point of the exercise.

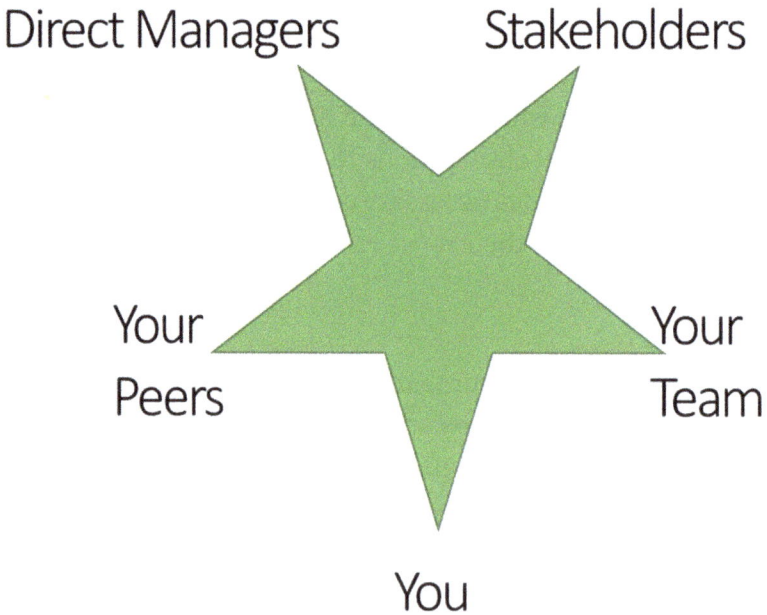

Figure 5 STAR model for potential audience discovery

Example of Discovery outcome

Conversation	Frequency	Audience	Priority
VP meeting	Monthly, usually first Wednesday of Month	Joe, Pradeep, Karthik, Amy, Mason	1
Project Status Meeting	Weekly on Monday	PM and PM Team	4
Finance Review	Monthly on first Tuesday of Month	Bill, Susan – Accounting, Joe, Pradeep, Karthik, Amy, Mason	2
3 + 9 Finance Exercise	Quarterly	Bill, Susan – Accounting, Joe, Pradeep, Karthik, Amy, Mason	3

Designing the context

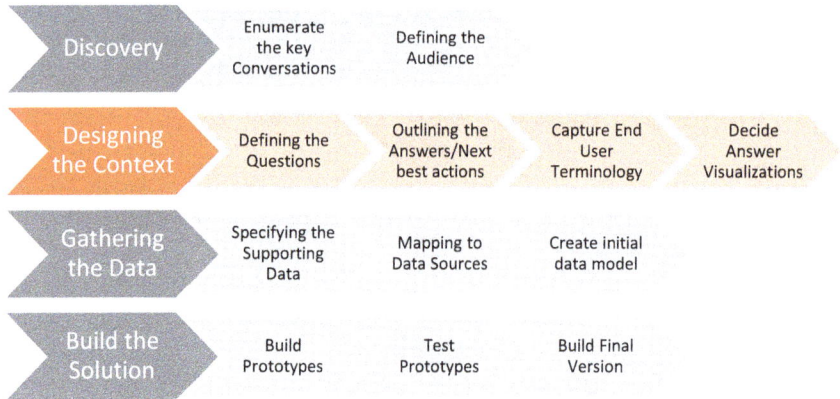

Once we have our prioritized list of conversations from Discovery, we start to break down the activity into our four areas. These areas are:

- Defining the questions
- Outlining the answers
- Capture end user terminology
- Decide answer visualizations

Defining the questions

Conversation-Centric Design starts with a key conversation. We'll discuss how to decide what those will be shortly. For example, let's use the Monthly VP Meeting as an example.

Using a meeting like this, we gain a lot of useful information quickly. First, we know what the conversation is about as there is an agenda for this meeting. We know the audience that will attend the meeting. Looking at the figure below, we can see that two items are already captured.

When you drill into the meeting agenda, rephrase the items as a series of key questions. For example, your list could look like this:

Original agenda item	Question version
Budget Update	Where are we on our overall budget? What are our top five deviations from budget?
Key Projects Update	What projects are running late? What projects are finishing this month? What projects are starting this month?
Personnel Update	What roles are short-staffed over the next six months? What roles are being filled primarily via vendors?

The next step is to group, prioritize, and triage the questions as it may not be possible to address all questions due to resource constraints or other factors. Optimally, this should be done by the attendees. They should rank the questions in importance and the input should be used to stack them.

Outlining the Answers

Developing the answers for each question requires deciding several aspects to ensure you have what you need. The goal is to uncover mismatches between requirements and process.

For example, one VP has decided that the Budget reports need to be refreshed daily. However, the process that delivers the budget data only runs monthly. Therefore, do you as the creator want to face a VP who has been looking at the reports daily expecting changes, or do you want to make them aware of this issue upfront?

Data

In examining the questions, a key need is finding the source and data elements needed to address the question. For example, the question "Where are we on our overall budget?" implies that the budget numbers are in a system that can be accessed and can be matched with expenditures in some fashion.

In some cases, this may be an Excel spreadsheet managed by one person. One side effect of this investigation is that other process changes may be warranted to support the desired BI need.

Tool

Microsoft offers three core tools for Business Intelligence:

- Power BI
- Power BI Paginated Reports
- Excel

These tools have distinctive features and capabilities that suit unique needs and scenarios. It is essential to pick the right tool for the need to avoid creating solutions that are fragile, complex, or inefficient.

For example, if you need to print a report to provide to field personnel, you may want to use Power BI Paginated Reports rather than Power BI itself. Power BI Paginated Reports are designed to create pixel-perfect, print-ready reports that can be easily exported to PDF, Word, or Excel formats.

Power BI, on the other hand, is more suitable for creating interactive and dynamic reports that can be viewed and explored online or on mobile devices.

By choosing the right tool for the need, you can ensure that your Business Intelligence solutions are effective, reliable, and user-friendly.

Definition

When formulating the answers, it is especially important to clearly define the conditions that influence or are used in filters, KPIs, measures, conditional columns, and other constructs. As discussed in Chapter 1, this "Tower of Babel" problem is common. Certifying key definitions are consistent is a particularly important aspect in ensuring expectations are met and all data consumers have the same shared understanding of the data represented.

Granularity

Once we figure out that the data exists and can be accessed, the next question is one of granularity. How granular does the answer need to ultimately be? This will create the basis for groupings, pivots, and what's the lowest level of data available to view.

This question should be asked because of the unintended consequences on organizations. This one question can significantly affect the amount of overhead work by an organization.

Case in point, we were doing a work study for a major consumer goods company to determine how a Project Manager (PM) spends their day. PMs were complaining about spending too much time dealing with reconciliation. As we investigated this complaint, we found that each PM was spending 8+ hours/week reconciling costs totals between the Project Management system and the Finance system. The CIO and other senior managers were puzzled by this. Further interviews with the PMs determined the effort was driven by two factors.

- All of the dollar amounts presented in reporting were "to the penny" level of granularity
- PMO process required that the Cost totals from both Project and Finance systems match

Consequently, PMs were spending a lot of time tracking down mismatches that were less than $10. In many cases, the total amount of discrepancy was much less than the cost of the PM for 8 hours. Also, the total Project portfolio was $680 Million dollars, so this level of precision was unnecessary.

I discussed the situation with the CFO, asking at what level of granularity should we be tracking to for this need? He said, anything below $10,000 at this portfolio size wasn't significant and approved us truncating all project cost totals to the $10,000 level.

The impact of this was felt at once by all PMs. Suddenly, the reconciliation work disappeared as numbers from both systems balanced 99.8% of the time at this level of granularity.

Next best action

One thing to capture as you outline the answers needed is what is the likely next action that will be taken when a condition is encountered? For example, if an indicator is red, a PMO manager may contact the PM directly over chat to ask questions. If it was yellow, they may include it as an agenda item in the next meeting. Making it easier for the user to execute the next best action directly from your content will make it more likely that the content continues to be used.

Process of data delivery, timing, and access

One of the aspects that we need to consider when designing BI content is the process of data delivery, timing, and access. This means understanding how the data is created, updated, and delivered to us, and how it affects the functionality and quality of our BI content.

The process can vary depending on the data source, the data type, and the business impact of the data. For example:

- Some data sources may allow us to access the data directly and in real time, such as databases, APIs, or web services. This enables us to have faster access and more up-to-date data for our BI content. However, this may also require more technical skills and security permissions to connect and query the data sources.
- Some data sources may only provide us with the data periodically and in a summary format, such as CSV files, Excel files, or PDF files. This may limit our access and delay our data for our BI content. However, this may also simplify our data processing and ensure our data quality and consistency. This is often the case for high business impact data such as financial information, which may have strict rules and regulations for data governance and compliance.
- Some data sources may have a combination of both direct access and periodic delivery, such as cloud platforms, online applications, or hybrid systems. This may offer us more flexibility and options for our BI content. However, this may also create more complexity and challenges for our data integration and management.

How the data is delivered and accessed can also affect the quality of our BI content. For example:

- If we have direct access to the data source, we need to ensure that the data is accurate, complete, and reliable. We need to check for any errors, outliers, or anomalies in the data. We also need to handle any changes or updates in the data source schema or structure.
- If we only have periodic delivery of the data source, we need to ensure that the data is relevant, timely, and consistent. We need to check for any gaps, delays, or discrepancies in the data. We also need to include the last modified date of the data in our BI content so that the users are aware of the data's age.

Therefore, when designing BI content, we need to understand the How the data is delivered and accessed for each data source. We need to choose the appropriate data source type, format, and frequency for our BI content requirements and objectives. We also need to consider the trade-offs between speed, functionality, and quality of the data.

Capturing End User Terminology

Power BI has a powerful ad hoc querying tool called Q&A. Q&A allows you to ask natural language questions about your data and get instant answers in the form of charts, tables, or cards. Q&A relies on a detailed linguistic schema to map the user's terminology to the data elements in the Power BI Model. The linguistic schema is a set of rules and definitions that tell Q&A how to interpret and process the questions. It includes synonyms, phrasings, terms, and expressions that the users use when referring to the data elements.

As you are developing and designing questions and answers, it is important to capture the language used by the users when referring to a data element. This can be incorporated into the model's linguistic schema, making Q&A much more useful and correct.

For example, if you have a column called ProjectName in the data source for the Project entity but your users call the entity "Gantts," "schedules," and "plans," you can add those three terms as synonyms for ProjectName in the linguistic schema. This way, Q&A will understand that when the consumers ask questions like "How many Gantts are behind schedule?" or "What are the total costs of the plans?" Q&A will know the user is referring to the ProjectName column.

Decide answer visualizations

Once you have the question and the answer with permutations decided, you can classify your answer to determine the best set of visualizations for the need. By permutations, the answer may require distinct groups, levels of rollup, etc.

I advocate the Data Visualization Literacy Framework, created by Indiana University as a great approach to understanding how to create the correct visuals for the identified question. The DVL–FW has seven types of visualizations.

Categorizing and clustering

- Categorization is the assignment of your data to a meaningful category of similar data.
- Categories may be manually defined or computed automatically.
- Clustering is a form of categorization where data is grouped so that the data in a cluster is more similar than the data in other clusters. Clustering is used to make patterns easier to find and reduce overall visual complexity
- Recommended Power BI visuals

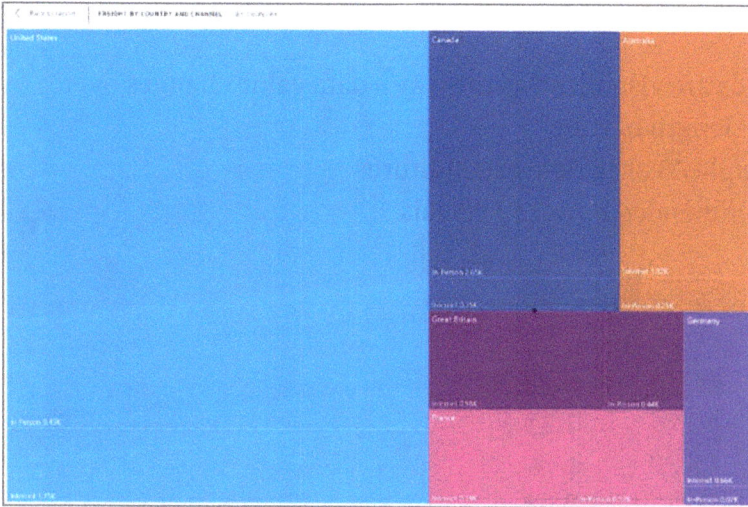

Figure 6 Categorizing using the Tree map visual

Trends - Time-Series

- Trends are visualized as the way a data value changes over some length of time
- Example: Year to date expenditures
- Recommended Power BI visuals

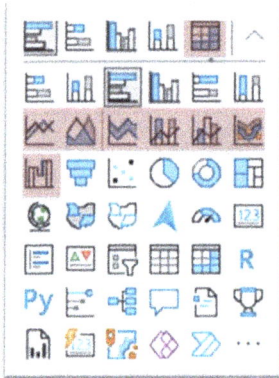

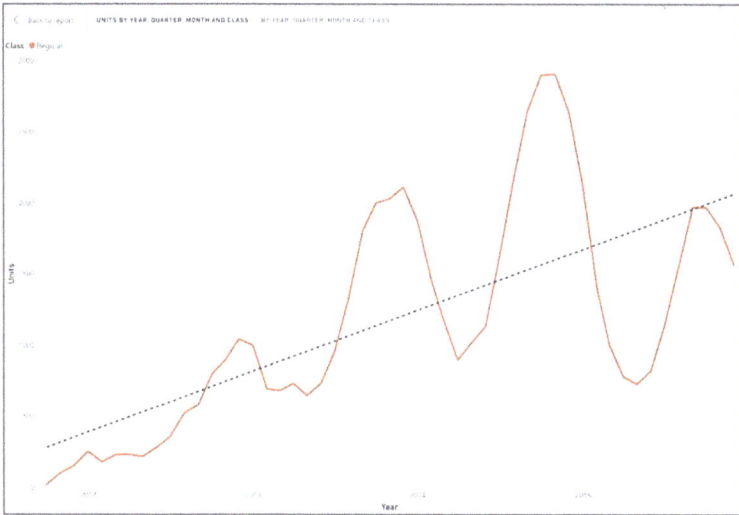

Figure 7 Visualizing a Trend using a line chart with trend line

Order, Rank, Sort

- Comparison of two or more values to illustrate relative contribution in a most to least or least to most pattern
- Example: Top 5 projects by total budget
- Recommended Power BI visuals

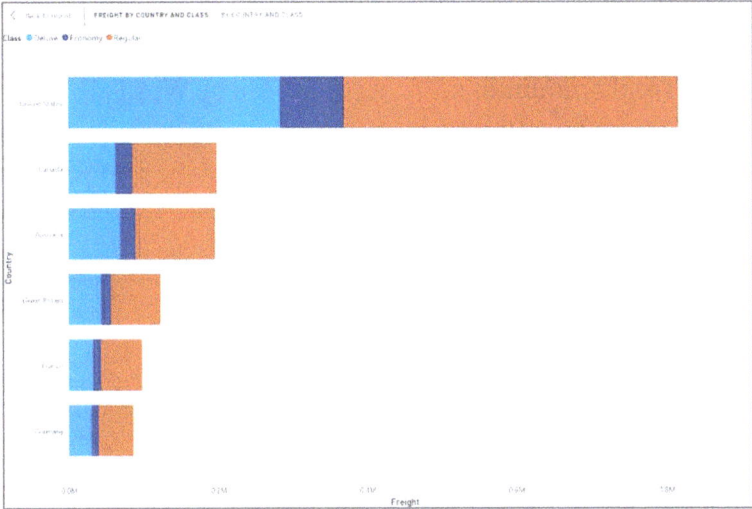

Figure 8 Showing a stack rank using a Stacked Bar Chart

Compositions – Part to Whole

- This refers to the way data elements are related to form a whole
- Example: Total amount of storage consumed by Word Documents in a SharePoint
- Example: Gantt chart visualization of tasks and their impact to the overall timeline
- Recommended Power BI visuals

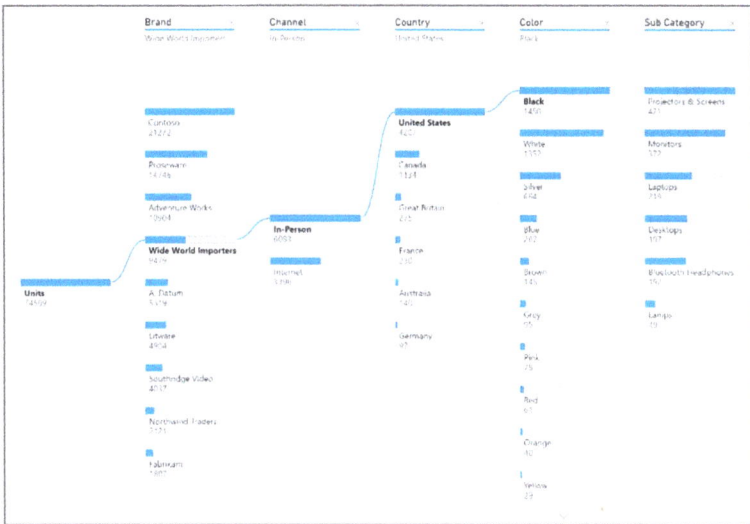

Figure 9 Showing decomposition using the Decomposition Tree visual

Distribution

- Visualization of data distribution by frequency of occurrence, either around a mean, timeframe or some other nominal value
- Example: Product demand by age group
- Recommended Power BI visuals

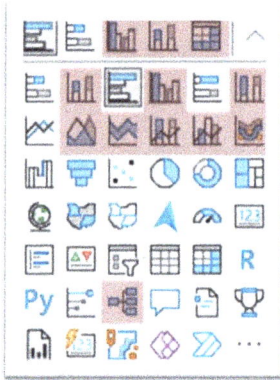

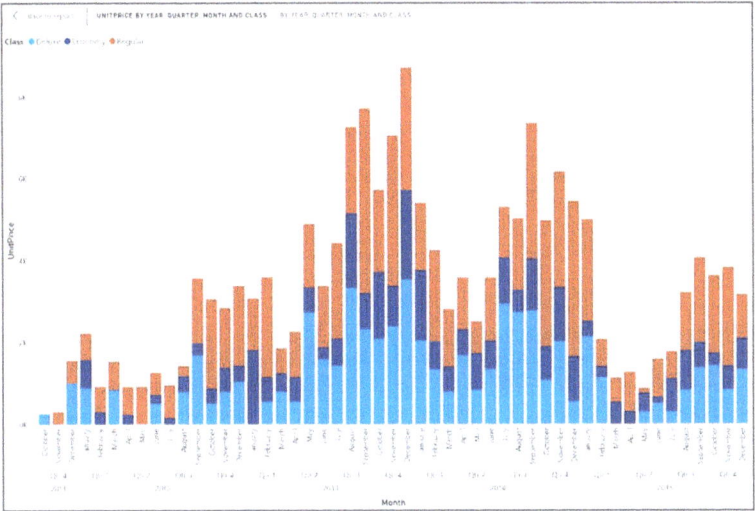

Figure 10 Showing distribution with a stacked column chart visual

Comparison – Deviation

- Comparison of data points against expected data values
- Example: Monthly actual sales versus projected sales
- Recommended Power BI visuals

Figure 11 Showing deviation with trend using KPI visual

Correlation/relationships

- Illustrates the relationship between two or more data elements
- Example: Job performance ratings versus social likeability scores
- Recommended Power BI visuals

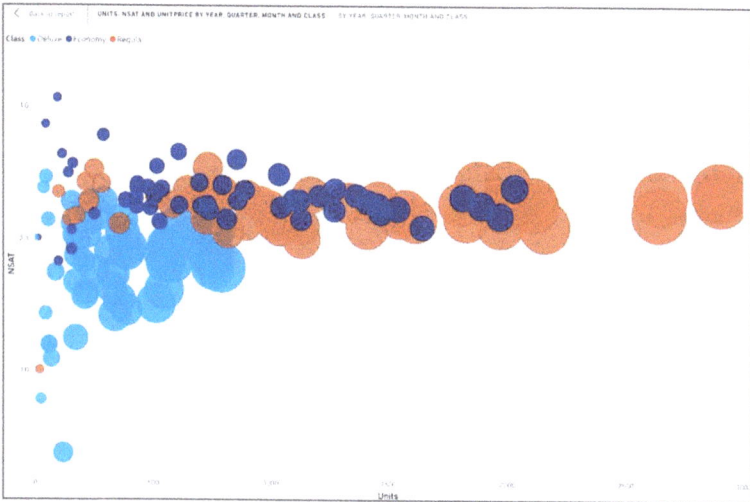

Figure 12 Showing relationships using the Scatter chart visual

Geospatial

- Shows the distribution and relationship of data values and elements in physical space
- Example: Total freight paid by nation for Summer 2023, worldwide.
- Recommended Power BI visuals

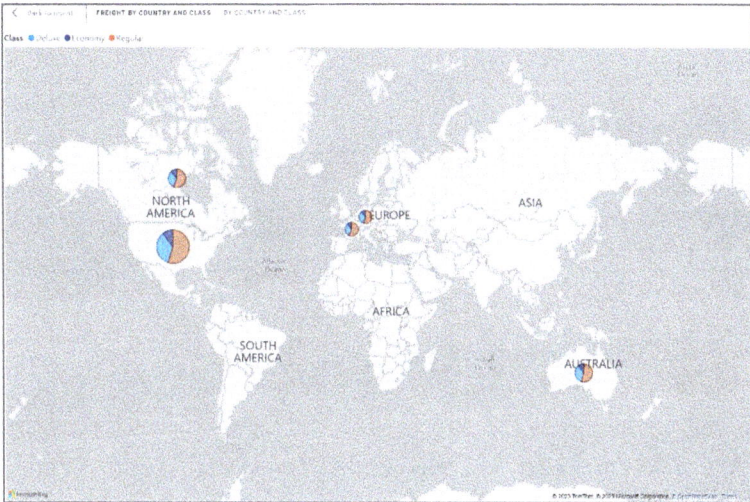

Figure 13 Showing freight charges by geographical location

Gathering the data

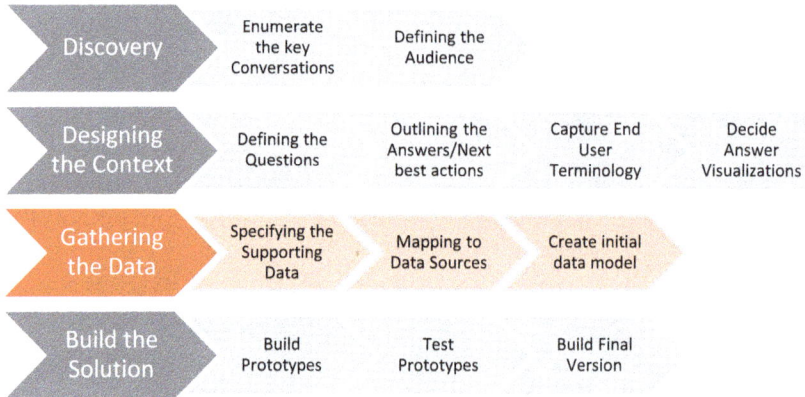

This phase involves the creation of the initial data model in Power BI or could be used to modify an existing data model. The goal is to ensure

- The list of data sources is known so that security access can be requested
- The requisite data is available from the identified data sources
- Create the initial data model

Specifying the supporting data

A review of the answers outlined above provides an opportunity to determine the source of the answer's data.

If your group owns the data, then follow your normal process to connect to the data.

If the data is coming from an external system source that is managed by another entity, access to the data source should be requested. This should also include the creation of data access accounts with requisite licensing. It is a good idea to be engaged with the data source's change management process so that as they upgrade and maintain their systems, your process is not broken by inadvertent changes.

If the data is being sourced from a manual creation process, there is a need to standardize the location and name of the requisite data files. A change management process for the data structures should also be considered for ongoing management. Doing so ensures that inadvertent changes don't break the data refresh going forward.

Mapping to data sources

Once all data sources have been identified, three aspects must be rationalized to ensure use of the data will lead to actionable data.

In many cases, you will need to combine data from the various sources to create a model. Doing so requires a set of common key values between tables. For example, if you are tracking a rolling total of expenditures and need to compare them to the budget, you'll need a key such as budget code on both the budget and expenditure records to tie them together.

In some cases, a multi-part key, involving multiple column values, must be used to denote a unique record. To ensure Power BI can work with this effectively, you'll need to synthesize a compound key, where all the values of the multiple columns are concatenated into a single column value.

Differences in data granularity is another difficulty encountered. Pulling data from various sources are usually at different levels and types of granularities. This creates issues rationalizing the data. For example, your budget information is only available by cost center and your expenditures are at the transaction level. How would you tie these together to create a project level summary?

Mismatches in data refresh cycles can also lead to integration issues. Some data may be real-time, others refreshed once a month. As you bring the data together, a time scale minimum granularity should be defined.

Create initial data model

A data model contains three primary elements that the CCD process should assist you in capturing. The first is the data needed to create business intelligence content. The second is metadata, by which we tell Power BI how to treat the data. For example, if the Next best action for a given visual is to navigate to the source system, we can create a custom URL using Power query or DAX as data. We then update the metadata to tell Power BI to treat this data as a clickable URL and display it as an icon.

Lastly, we incorporate the business rules that must be in place to curate and interpret the data. In an earlier story, we mentioned how the same Project Health indicator was defined differently by each CVP's group. Centralizing business rules and definitions into shared datasets will help prevent definitional drift of metrics.

A data model is ideally structured in a star schema pattern. By this, a model generally has facts and dimensions.

Facts are the events that have occurred. Examples of this can be transactions, timesheet entries, status updates, etc. Dimensions are those entities to which the fact applies. For example, for a transaction, dimensions could include billing company, cost center, project, etc.

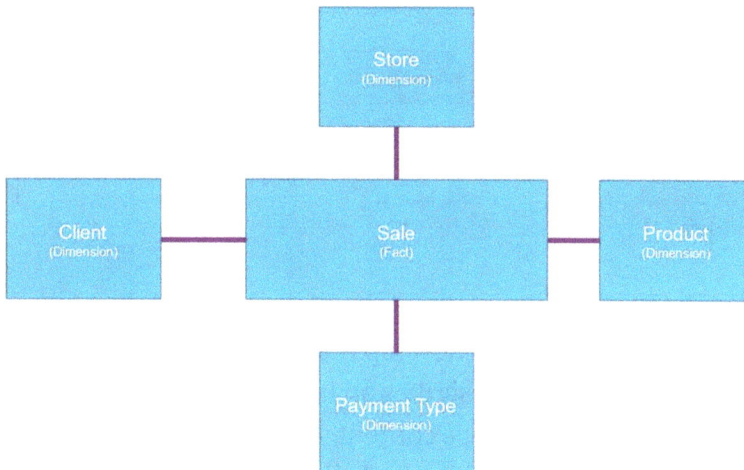

Figure 14 Example of star schema

Once the core data is in place and structured correctly, the initial measures, summary tables and other transformations should be put in place. In many cases, a company may have standard elements that are created for a dimension and/or a fact table. These could include a date table, for example. This creates the basis for the next phase, where iterative development is used to finalize the model and requisite reports.

Build the solution

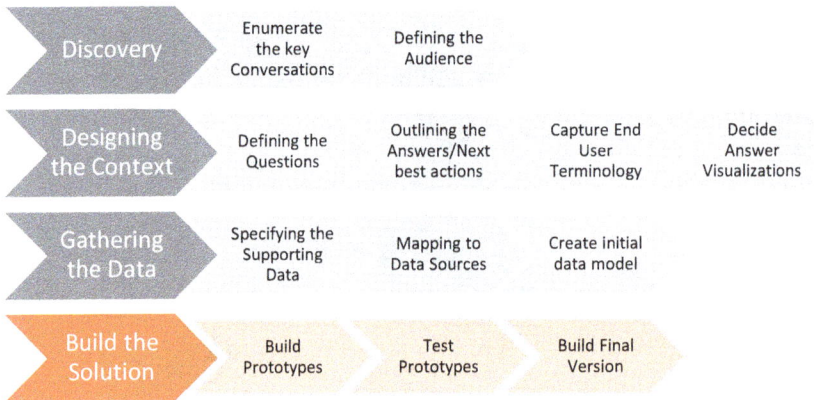

Discovery	Enumerate the key Conversations	Defining the Audience		
Designing the Context	Defining the Questions	Outlining the Answers/Next best actions	Capture End User Terminology	Decide Answer Visualizations
Gathering the Data	Specifying the Supporting Data	Mapping to Data Sources	Create initial data model	
Build the Solution	Build Prototypes	Test Prototypes	Build Final Version	

The building of the solution starts once you have a solid design. This should reduce the time to develop the solution and reduce any rework. Note, the following describes an iterative approach to developing business intelligence content. Many data consumers do not have the training or experience to formulate the final visualizations. The iterative process gives them a voice in the process as well as builds champions for the final implemented product.

Note, the process below is designed for large business intelligence projects with high visibility in the organization. The process can be streamlined for smaller projects as needed. The key learning is to drive the test process rather than simply reacting to it. Simply reacting leads to scope creep, unmet expectations and generally longer timelines to deliver.

Build Prototypes

The initial report prototypes validate the design process findings and are the first opportunity for the data consumers to be hands on. A focus group of data consumers should be identified from the earlier audience identification and recruited to help with this validation.

In most cases, the data author should be listening only, with the session facilitated by another with no vested interest in the feedback. Any feedback that is gathered should be reviewed, triaged as to whether it is of value to address, and then work assigned to incorporate feedback.

We use a PowerPoint-based prototyping toolkit with our clients to help them quickly create prototypes for review.

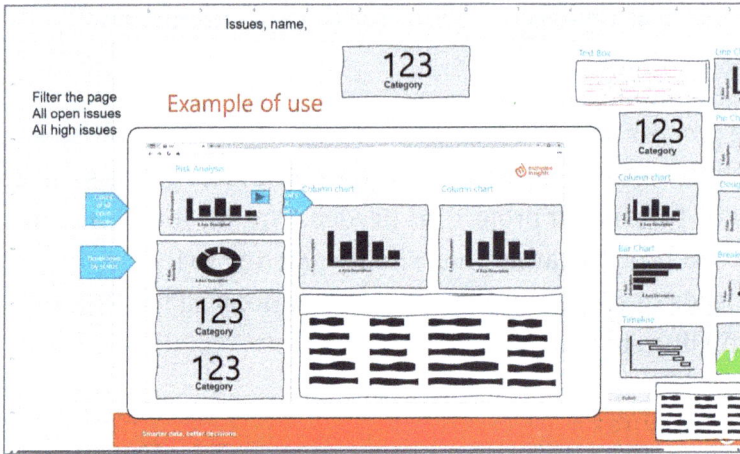

Figure 15 Marquee Insights prototyping toolkit

To prevent runaway scope, there should be a set number of review rounds initially and a data consumer representative should be part of the feedback triage process.

Test Prototypes

Once you've reached a specific level of quality with the model and reports, an extended pilot group should be given access to it to test the new content.

It is imperative that you provide clear guidelines to the new pilot group on how to file feedback, what is considered a bug, what is considered a change, and what will be done if it's simply product behavior. There should also be a task plan with clearly defined assignments. Otherwise, you'll have a new group of people wandering around in the new content with no context and focusing on product behavior. This will yield feedback of little value.

Typically, a few rounds are necessary to get everyone's feedback and to have the feedback triaged and prioritized.

Build Final Version

Once the test phase is over, the final build out of the content occurs. The work plan of remaining items should be tracked. The content should be packaged as an app and released according to internal procedures to the target audience defined earlier.

CHAPTER 3. A REAL WORLD EXAMPLE

Steve, a business analyst in the IT group, received the following request from Donna, an IT Director.

"Steve, we need a dashboard for our IT Managers and Finance personnel to use in the Monthly Review meeting. This will be used to see project related data. How soon do you think you can get that done?"

Steve sighed and thought, "It's time to put this CCD training to work."

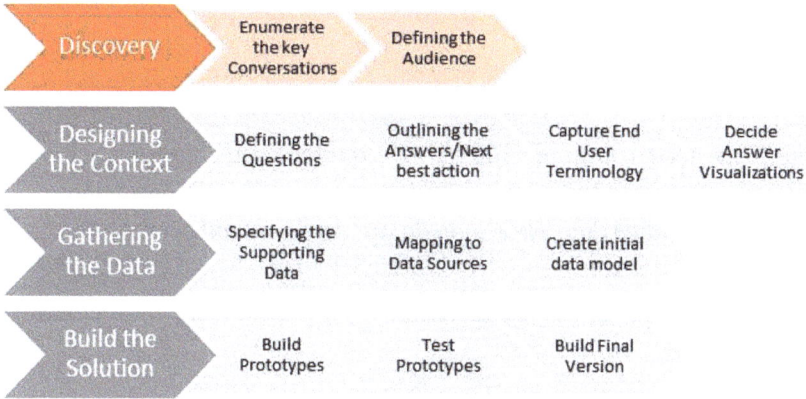

Discovery	Enumerate the key Conversations	Defining the Audience		
Designing the Context	Defining the Questions	Outlining the Answers/Next best action	Capture End User Terminology	Decide Answer Visualizations
Gathering the Data	Specifying the Supporting Data	Mapping to Data Sources	Create initial data model	
Build the Solution	Build Prototypes	Test Prototypes	Build Final Version	

Step 1. Enumerate the key conversations

First step is to enumerate the conversation. He highlighted the phrase as follows, as this gave him the conversation and the timing.

Steve, we need a dashboard for our IT Managers and Finance personnel to use in the Monthly Review meeting. This will be used to see project related data. How soon do you think you can get that done?"

Step 2. Define the audience

Next step was to determine the audiences. Steve highlighted the following.

Steve, we need a dashboard for our IT Managers and Finance personnel to use in the Monthly Review meeting. This will be used to see project related data. How soon do you think you can get that done?"

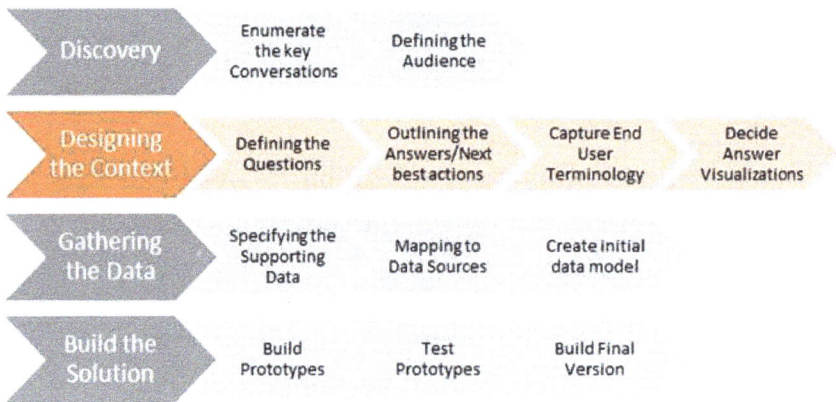

Step 3. Define the questions

Steve frowned, "What I don't see is any actual questions in this request."

He replied to Donna with some questions about what questions needed to be answered for the IT Managers and for Finance. This dialog continued for a bit.

As he learned in his training, he needed short questions that started with Who, What, When, Where, Which or How much. At the end of the exchange, he had the following questions.

Audience	Monthly Review Conversation
IT Management	Which projects are behind? Which projects are finishing this month? What are the upcoming key milestones this month?
Finance for IT	What projects are over budget? What are the expected capital expenses for this month?

Steve decided to focus on the first question to train Donna in what he needed, as this was her first time asking for content. Once she understood what he needed, the process would go quickly.

Since this was an extension to an existing model, he knew the data would be coming from Project Online. However, he still needed to identify the data fields once this request was clearly defined. He looked at Project and saw there were many projects. Also, what did they mean by behind?

He replied with the following questions.

With regards to the "What projects are behind" question,

What kind of projects should be included in the report? Capital? All?

Define the criteria for "behind. "

Is it Planned > Baseline?

Is it some percentage variance?

Is it schedule or cost?

After several iterations, they finally arrived at the following criteria.

- Show all capital projects over $250,000 in total spend where the planned finish date is greater than 30 days past their baseline finish date.

Step 4. Outline the answers/next best actions

Steve started breaking down what was needed in each answer for the questions. For example, for the question, "Which projects are behind?" required the **project name** and the **project manager name** so that the report user knew which project manager to contact and which project to discuss. The report user also needed the **project health indicators for budget, resources, and schedule,** to determine if there were greater project issues afoot. The **original schedule date,** the **current schedule date**, and the **date variance** were also required to gauge the magnitude of the schedule problem. He also discovered that the audiences tended to look at the information by a specific **Program** or for a specific **Project Manager**.

Steve asked Donna, "What normally happens when a project is behind?" She replied, "We will look at the schedule in Project Online and email the PM is we have questions." Steve then added URL to project in Project Online and Project Manager email as data elements needed for implementing Next Best Action navigation.

Step 5. Capture end user terminology

Steve noticed that IT referred to projects as **"schedules,"** whereas Finance called them **"Gantts."** As Steve hoped to use Power Q&A with this reporting, he noted these synonyms for the Project entity. He also noted that **CapEx** was used as an abbreviation for

Capital Expense. CapEx would become a synonym for the Capital Expense column.

Step 6. Decide answer visualizations

Steve gave some thought as to how this might be visualized. He decided that three levels of data were appropriate. This would provide several pivots without being overwhelming. This would also set up the report to enable drill through to other detailed reports.

He decides to use cards for the max delay information since it's an overall number. He'll use the Stacked bar chart visual to do the project stack rank by delay visual. Finally, he's using the table visual to show key information for each individual project.

His first prototype looks like the following. The entire page is structured around a single question and the number of visuals is kept to a minimum. He repeats this process for each of the questions.

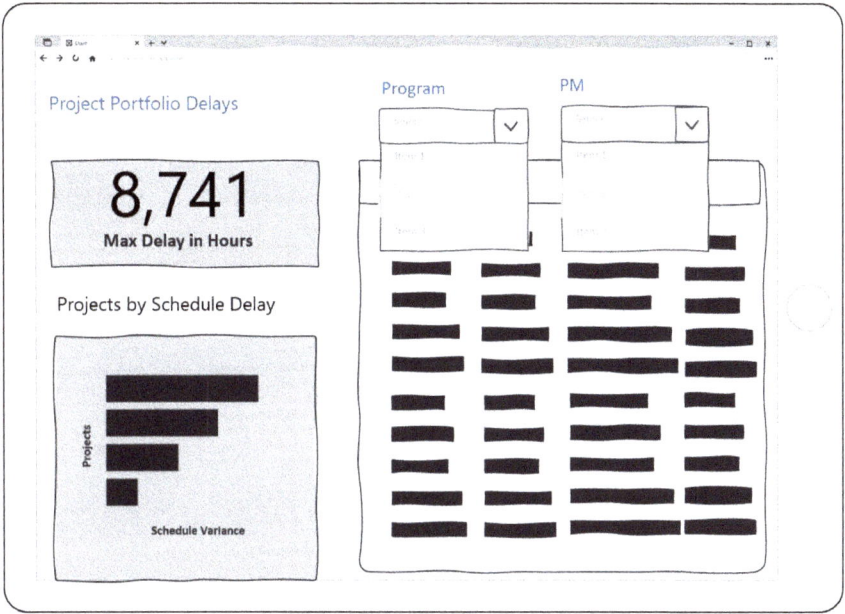

Figure 16 Steve's initial report prototype

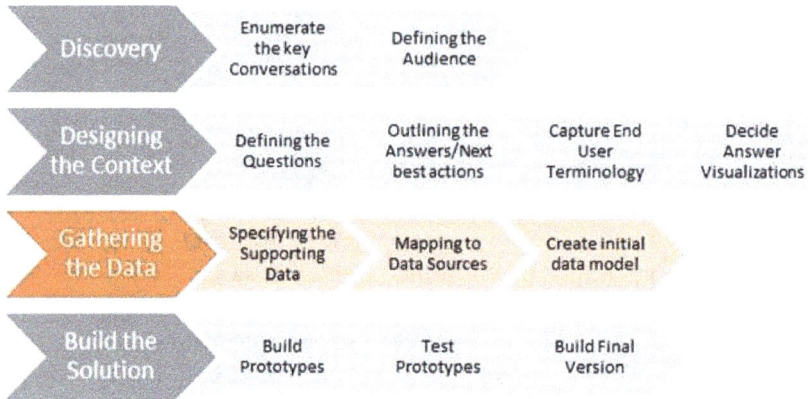

Step 7. Specify supporting data

Steve had accumulated the following logical columns that would be needed for this initial report to support calculations and filtering. This list would be used to determine later if there was special logic necessary to get any of this information out of Project Online.

- Project name
- Project manager name
- Project manager email
- Project budget health indicator
- Project resources health indicator
- Project schedule health indicator
- Program name
- Project finish date

- Project baseline finish data
- Project schedule variance
- Project cost
- Project Id

Step 8. Mapping to data sources

Steve's company had bought a pre-built Project Online Power BI solution from Marquee Insights. This made it easy to see what data was available.

Three data tables were required to get the above data: Project, Tasks, and Task Baseline. Most of the information was in the Project table but the challenge was getting the project baseline information in Power Query.

The project baseline information is stored in the Task Baseline table. However, to get the correct record for each project in Project Online, you have to merge Task Baseline with the Tasks table so that you can filter the results correctly for the Project Summary task. Steve will filter the Tasks table where TaskIsProjectSummary = True then merge it with the Task Baseline table where BaselineNumber is 0. Steve is really happy he took a Power Query class as well!

Project schedule variance is also stored in hours so he'll divide that by 8 so that he can show schedule variance in days.

Step 9. Create the initial data model

Steve creates a model with the following entities and relationships in Power BI Desktop.

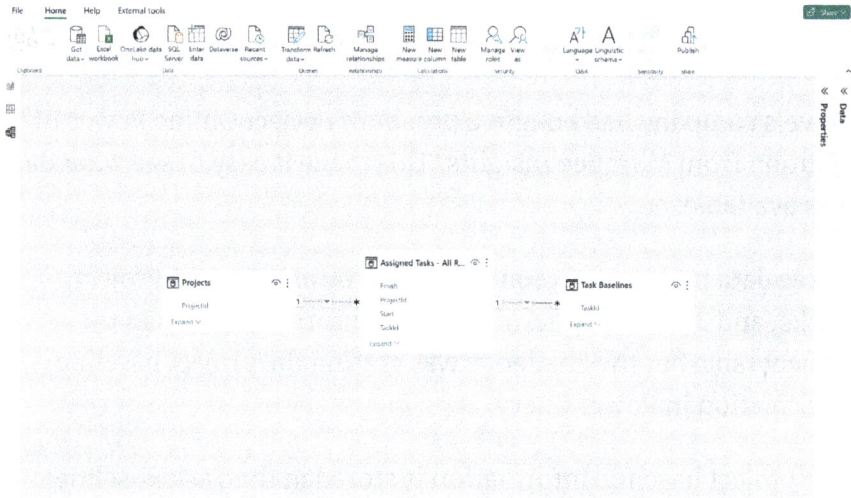

Figure 17 Steve's Power BI Model

Discovery	Enumerate the key Conversations	Defining the Audience		
Designing the Context	Defining the Questions	Outlining the Answers/Next best actions	Capture End User Terminology	Decide Answer Visualizations
Gathering the Data	Specifying the Supporting Data	Mapping to Data Sources	Create initial data model	
Build the Solution	Build Prototypes	Test Prototypes	Build Final Version	

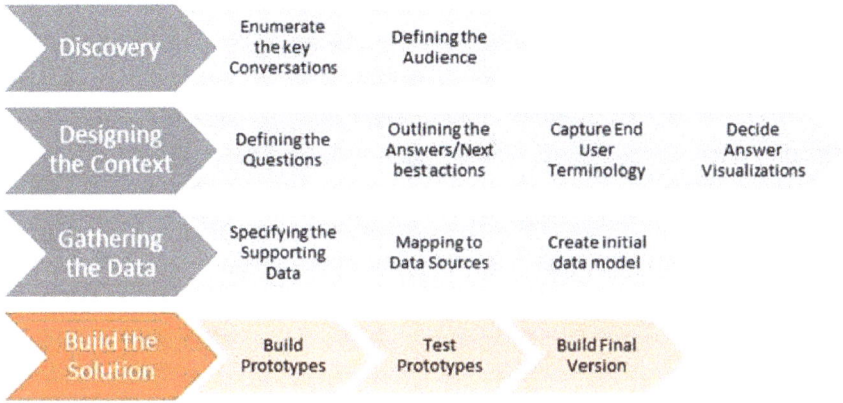

Step 10/11. Build/Test prototypes

Steve reviews the prototype with Donna. She sees how the reports are question focused and understands how these can be easily used and explained. She asks for specific drill throughs to other reports which Steve can do. She also appreciates the ability to add new pages as new questions arise, without having to re-architect every report.

Step 12. Build final version

Steve built the final version as shown below.

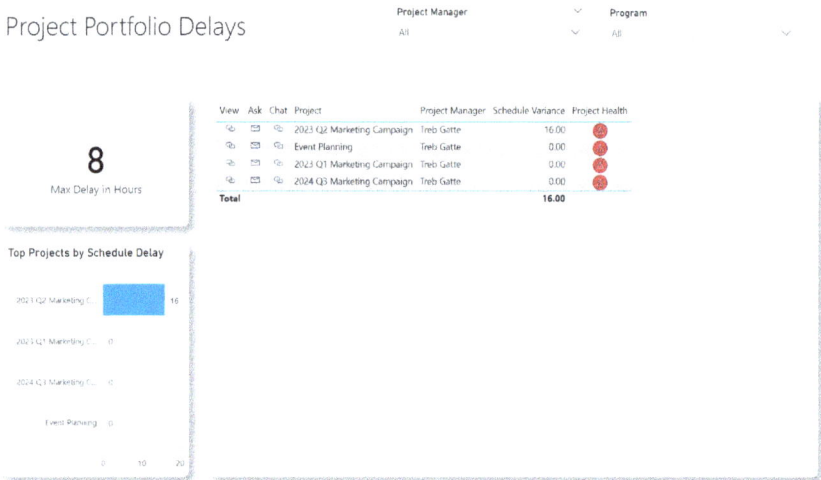

Figure 18 Steve's final report

Once Steve is done with the changes, Donna signs off on it.

He publishes the final version to a new workspace in PowerBI.com. This is used to create the **IT Monthly Meeting** app; which Steve publishes and assigns security rights to the IT Managers and Finance.

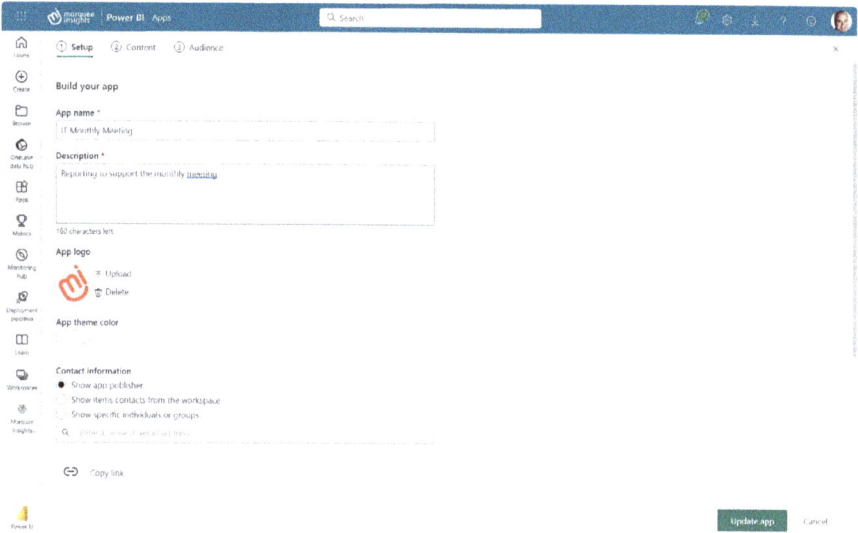

Figure 19 Steve's Report App

CHAPTER 4. SUMMARY

The Conversation-Centric Design™ process streamlines the development of business intelligence data by providing a framework that helps you anticipate and address issues before they arise in the development process. Once a solid base of content is in place, it also makes it easier to maintain that content over time as the requisite documentation is created as the content is developed.

Real-world use also shows it increases adoption of content since the delivered apps are aligned with the business activity that drove the need for the content. It also reduces the risk of changing reports since the exposure of a given report collection is limited to a single business activity.

For questions about Conversation-Centric Design™, contact us at hello@marqueeinsights.com.

CHAPTER 5. ADDENDA

The following related articles may be of interest to you as you continue your journey to creating a data culture. Each article covers a specific topic to Power BI.

These articles are intended to show you a different approach to Business Intelligence. After twenty years of data analytics, many companies are stuck getting their insights from manually maintained spreadsheets.

The two topics we'll cover impact how you implement the business intelligence content you create.

The first article addresses the difference between Business Intelligence and reporting. Many organizations fall into the trap of migrating existing views to a new tool, without taking advantage of the new tool capabilities. You should look more broadly as to how the tool can provide new capabilities.

The second article addresses our distrust of our own data and how this impacts acceptance of new solutions. It'll take you through some techniques that you can use to address this major implementation issue.

Business Intelligence is not just reporting

Whenever I start a conference presentation, I lead with "Business Intelligence is not just reporting!" This comes about as a call to get new Power BI users to look more broadly. If all you are doing with Power BI is replicating the Excel spreadsheets you already have, you are missing out on significant value. If your users want their new dashboards to look like a spreadsheet, a conversation is warranted.

Let's discuss three ways that Business Intelligence is so much more than reporting.

BI is your first step towards AI

A lot of companies today are intrigued by the potential of artificial intelligence. Some of them have started projects to implement AI. The issue that many companies will run into as they start the path toward AI is data quality. For AI to be successful, you must have good quality data to train the models. If you haven't started down the BI path yet, chances are, your data quality is poor at best. Jumping into AI with poor data quality will lead to incorrect outcomes.

BI forces you to clean up your data and processes

Another positive outcome of BI projects is a rationalization of entity and business rule definitions. For example, in a recent project, the client wanted to monitor product profitability across the board. The issue was that the definition of a product was different based on the division producing it. Consequently, we had to find common data ground for such an analysis to be made.

In other projects, we encounter timing discrepancies, dependencies on manual effort for key data, and incomplete data. It's very difficult to have useful near-real-time analysis over data that is received once a month from Finance. These timing issues must be worked out across processes.

BI can eliminate a lot of hidden manual effort

Manual data gardening of spreadsheets is taking up a lot of time in organizations and creating potential problems. Do you really want to bet your career on the numbers within a manually maintained spreadsheet? The sad truth is that many managers do just that.

Also, the manual effort is robbing you of your most precious resource, time. There was one organization where every one of their twenty project managers spent six hours a week manually creating status reports. Products like Power BI and Flow can be used together to automate the collection and dissemination of data within the organization, freeing up time for more valuable work. A small project like this can easily justify the license expense of Power BI.

BI can set the stage for further successes

In the end, a successful BI journey creates fertile ground for a potentially successful AI implementation. In the end, we will achieve Collaborative Intelligence (CI) where the AI tools augment the human and make them more effective. Many of the most amazing uses of AI are focused on shortening or removing the learning curve, instead of replacing the human.

BI can't do it alone

Lastly, one of the other challenges we see is the idea that simply bringing in a new BI tool will magically produce results. Time and again, organizations make software investments without the requisite investment in people. To get the highest benefit from a BI tool, your organizational culture must make data part of their day-to-day activity. Creating a data culture requires active investment and time.

The rise of Data Culture

One sign of a company that has achieved a data culture is that data has a "voice." You'll hear people ask in meetings, "what does the data say?" If you are starting the journey, you should consider how you will invest in training and in a community of practice to sustain the change long term.

In Data We Trust?

I finished the demo of Microsoft Power BI dashboards and reports that we had built for a client. I looked at the room and asked, "What do you think?" This proof of concept hopefully created excitement, by showing what was possible with the client's data. As we went around the table, people were generally thrilled. The last person's feedback, however, caught me off guard. "It looks great, but how do I know I can trust the data?"

That question rattled around my brain for days. In the rush toward a data-centric future, clients weren't asking if their data was trustworthy. I researched the problem further, finding some whitepapers and such on the topic, but no clear recommendations on how to address this issue.

Three Areas of Data Trust Issues

Data Authenticity

Is the data you are using authentic, in that is it from a trusted official data source? How do you know? Imagine your executives making decisions about your project portfolio based on a manually maintained spreadsheet instead of from data retrieved directly from their Project Management software.

A trusted data source is one aspect to consider. You also need to know if that data source is actively managed? It's one thing to have data from an official source but if it's a one-time extract versus an ongoing process, the value declines quickly.

Lastly, is this data coming from the official system of record? Imagine getting project cost values from a non-accounting system? Is this system the official system of record for cost data? Many reporting solutions obscure the source of their data, making it impossible for end users to determine if the source is the correct and official authoritative.

Data Integrity

Data integrity is knowing that the agreed upon business rules within your organization are consistently applied to your data. Integrity also looks at how close you are pulling data from the official data source. Is the data being taken as it is from the official data source or is it being derived? If it is derived, does it officially defined internal business rules to achieve the outcome?

For example, you are using Project Health from your project management system. Is the value following the standard Project Management Office definition for Project Health or is it using some specialized logic that maybe was used for a one-time analysis? How do you know? Deriving data is not bad if the process adheres to the established internal rules but you must have a way to gauge this.

Data Timeliness

When you go to the grocery store to buy fresh fruit or meat, the ability to assess food freshness is vital to your buying decision. Old fruit isn't very appealing and can have detrimental health effects. The same could be said about old data.

In data, we should be going through a similar assessment of freshness. Is this data representative of recent activity? This is not when was the data last refreshed into the report, but rather when was the data last modified. Data that has been modified this morning is likely to be more representative than data that was modified three weeks ago. How do you gauge the freshness of your data?

Authenticity revisited

Next, let's explore the difficulty in determining whether your business intelligence content is using authentic data. To illustrate the point, let's examine a recent Seattle Times article about the Measles outbreak happening in Washington.

The article in question, "Are measles a risk at your kid's school? Explore vaccination-exemption data with our new tool," presents a story filled with data charts and tables and made some conclusions about the situation. Many internal reports and dashboards do the same, presenting data and conclusions. Unlike internal reports, newspapers list the source and assumptions in small print at the end of the story. Knowing the data comes from an official source adds authenticity.

The following note is supposed to increase authenticity.

"Note: Schools with fewer than 10 students were excluded. Schools that hadn't reported their vaccination data to the Department of Health were also excluded.

Source: Washington Department of Health (2017-18)"

But did it really? Documenting any exclusions and note sources is a good practice. However, it's not very prominent and if you search for this data, you'll likely find several links. There's no link or contact information.

Data authenticity is crucial to making successful decisions. To do so, key data questions should be answered.

What data was used?

Many content creators don't bother to document the source of their information. Many would not have the same level of confidence about the new financial dashboard if the viewer knew the data came from a manually manipulated spreadsheet, instead of directly from the finance system. How would the reader know anyway? In many cases, they wouldn't. The Seattle Times provided a hint, but more is needed.

When you buy items like wine, you know what you are buying because the label spells it out. A wine bottle is required to have a label with standard data elements to ensure we know what we are buying. For example, a US wine label must have the type of grape used to make the wine. Red blends must list the varietal and percentage so that the customer is clear on what is in the bottle. Having the equivalent type of labeling would improve transparency about data authenticity.

Who owns the data we are consuming?

This is very important, especially if we spot incorrect or missing data. Who do we contact? The Seattle Times lists the Washington Department of Health as the data owner. This is a good starting point but doesn't completely fill the need. For internal reports, all data sources should include an owning team name and a contact email. The data vintage page should also include the site URLs and a contact email for the data sources.

How old is the data?

The age of the data in your data source strongly influences whether it can be used for decision making. By default, we know the last time the data was refreshed. However, that does not address the client's need. We need to evaluate whether the data has been updated within a client's acceptable timeframe as determined by their business rules. In our Marquee™ products, we include a data freshness indicator that shows proportionally how much of the data has been updated recently. Recently it has become a business rule of what constitutes fresh data. With some companies, the entity must have been updated with in the last seven days to be considered fresh.

How to create a data vintage page?

The following steps enable you to create a simple "data vintage" page in your Power BI model.

- Create a Data Vintage page (you may need more than one, depending on how many datasets and sources you have)
- Add a back button to the page. We put ours in the upper left corner
- Add the following information to the page using a consistent format that you've decided upon
  - Name of dataset
  - From where the data is sourced and how often
  - Which team owns the data
  - How to contact the data owner, if available

- Create a Data Vintage bookmark for the data vintage page so that it can be navigated to via a button.
- Go back to the report page that you created from this data
- Add an Information button to the upper right corner of the page.
- Select the button and navigate to the Visualization blade
- Turn on Action
- Set Type to Bookmark
- Set the Bookmark to the one you created in Step 4.
- Ctrl + Click the Information button to test
- Ctrl + Click the Back button to test

Anytime a user or fellow Power BI Author has a question about the underlying model data, it can be accessed very easily. You'll also improve impressions of data authenticity by implementing this label in a consistent manner across all content.

Author Bio and Company Profile

Trebuel Gatte, CEO, MarqueeInsights.com

With 24 years of experience, Treb Gatte is a widely recognized business intelligence expert.

Prior to becoming CEO, he worked in leadership positions at Microsoft, Starbucks, Wachovia (now Wells Fargo). He has been recognized for many years as Data Platform Most Valuable Professional by Microsoft.

About Marquee Insights

Founded in 2013, Marquee Insights serves our Fortune 500 clients worldwide. With expertise in healthcare, financial services, retail and engineering industries – we are dedicated to enabling better decision making.

Our passion is to grow your data culture with the human touch – we recognize people, processes and tools are all needed for success long term.

Based in Bellevue, Washington, USA, near to Seattle and close to Microsoft and Amazon's headquarters, our expert team is well versed in the challenges and opportunities in business intelligence. Our Austin, TX office is also located in a national tech center. These locations give us access to a great talent pool.

Marquee Insights' parent company, Tumble Road LLC, is a Microsoft Certified ISV Partner. Contact us at hello@marqueeinsights.com

If you need ready-made Business Intelligence solutions for Microsoft Project, Microsoft Office 365, Jira, and more, please see our offerings at https://marqueeinsights.com

marquee
insights™
Smarter data, better decisions.

FOLLOW ON SOCIAL MEDIA

Blog

https://marqueeinsights.com/blog

LinkedIn

https://www.linkedin.com/company/marquee-insights/

X/Twitter

https://twitter.com/marqueeinsights

www.ingramcontent.com/pod-product-compliance
Lightning Source LLC
Chambersburg PA
CBHW070942210326
41520CB00021B/7018